History and Physical Exam Documentation Manual: A guide for medical students entering Core Clinical Rotations:

26 Clinical Cases Reviewed for Internal Medicine, Surgery, Pediatrics, Psychiatry, and Obstetrics & Gynecology.

BY: Y. Ali M.D.

Preface:

During my years as a medical student first, then as a resident physician, and now a practicing physician; I have found that many medical students making the transition from a second to a third year have difficulties with documentation of their findings with patient encounter. Inspired by them, I write this book to share some examples on how to document during core clinical rotations in third and fourth year of medical school.

My attempt is no way to undermine previous works, nor to serve as a sole guide on documentation but a humble attempt to share few examples that I hope would inspire young minds in pursuit of excellence. Furthermore, every physician has their own way of documenting their findings. I sincerely hope that this book would help the readers solidify basics and help them develop their own style throughout many years to come.

Lastly, I dedicate this book to everyone in my life that has helped me accomplish my goal of becoming a physician; my teachers who taught me, my mentors who advised me, my family members, and loved ones. Without their love that I treasure very much so and fortunate support throughout the years, I would not have had the opportunity to serve my patients and pursue what I feel passionate in life; which is the practice of medicine.

- Yours Truly,
Y. Ali M.D.

Disclaimer:

This book is not intended as a substitute for the medical advice of physicians. The reader should regularly consult a physician in matters relating to his or her health and particularly with respect to any symptoms that may require diagnosis or medical attention. All rights are reserved. It may not be reproduced without the permission of the author of the book in any form. Furthermore, the literary work itself does not pertain to any single person living or dead but a compilation of several patient encounters in the clinical setting and any complete similarities are purely coincidental in nature.

CLINICAL CASES OUTLINE:

.......Preface………………………………………………………………….. 2

1. Internal Medicine:

.......Case Study 1…..Alzheimer's disease……………………………….. 9
.......Case Study 2…..Asthma………………………………………………. 11
.......Case Study 3…..Congestive Heart Failure ………………………….. 13
.......Case Study 4…..Generalized Fatigue……………………………….. 15
.......Case Study 5…..Gout…………………………………………………. 17
.......Case Study 6…..Hematochezia……………………………………… 19
.......Case Study 7…..Myocardial Infarction……………………………… 21
.......Case Study 8…..Pneumonia ………………………………………… 23
.......Case Study 9…..Rhabdomyolysis…………………………………… 25
.......Case Study 10….Transient Ischemic Attack ……………………….. 27
.......Case Study 11….Gastrointestinal bleeding ………………………... 29
.......Case Study 12….Hematuria …………………………………………. 31
.......Case Study 13….Asthma……………………………………………… 33

2. Surgery:

.......Case Study 1…..Acute Cholecystitis………………………………….. 37
.......Case Study 2…..Acute Appendicitis………………………………….. 39
.......Case Study 3…..Inguinal Hernia……………………………………….. 41
.......Case Study 4…..Hyperthyroidism……………………………………… 43
.......Case Study 5…..Hemorrhoids………………………………………….. 45
.......Case Study 6......Renal Cell Carcinoma……………………………….. 47
.......Case Study 7…..Appendicitis…………………………………………… 49

3. Pediatrics:

.......Case Study 1…Otitis Media…………………………………………… 53
.......Case Study 2…Strep Throat………………………………………….. 55

4. Psychiatry:

.......Case Study 1….Depression…………………………………………… 59
.......Case Study 2….Schizophrenia ………………………………………… 61
.......Case Study 3….Bipolar Disorder……………………………………… 63

6. Obstetrics & Gynecology:

……Case Study 1…..Preeclampsia………………………………………..   **67**

……Abbreviations ……………………………………………………   **69**

Internal Medicine

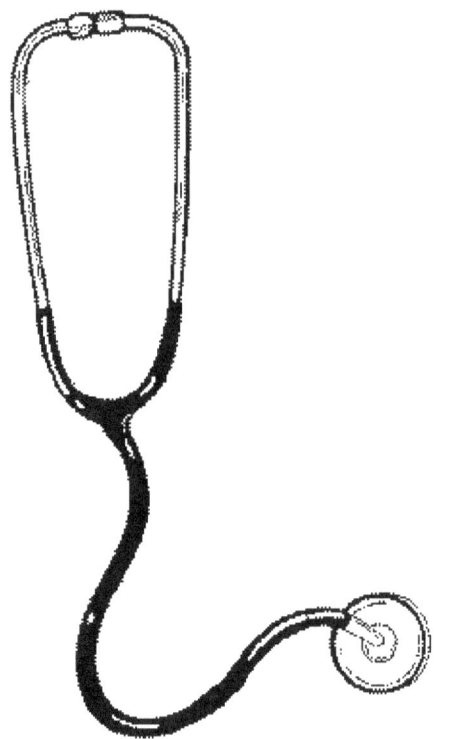

CASE STUDY 1: Alzheimer's disease

Chief Complaint: Memory loss and forgetfulness.

History of Present Illness:

Patient is an 82 year old Caucasian male who is brought in by his daughter to the family physician. The daughter complains that her father has been losing his memory for the past 2 years. He seems to forget where he puts his "stuff" such as watch, clothes, shoes, etc. Patient has occasionally also forgot the directions to come home from the grocery store. Patient is not fond of doctors and has not seen a doctor for approximately 10 years in an outpatient setting. He does admit to occasionally having problems urinating, increased number of times using the bathroom, while dribbling seen at other times. Patient denies any other problems. Patient denies any nausea, vomiting, diarrhea, constipation, shortness of breath, chest pain, and abdominal pain. The daughter also states that her father seems to use the bathroom a lot. She is often wakening up at night from her father going to use the bathroom.

Past Medical History: Hypertension and Dyslipidemia.
Past Surgical History: none
Family History: father and mother both had a history of HTN and DM II.
Medications: Metoprolol, Metformin, Glipizide, and Acetaminophen for pain.
Allergies: NKDA
Social History: no smoking, no ETOH, no drugs, single lives with his daughter after his wife passed away.

Review of Systems:

General: negative
Skin: negative
HEENT: negative
Neck: negative
Chest: negative
CVS: negative
Gastrointestinal: negative
Urinary: increased urinary frequency and dribbling.
Musculoskeletal: negative
Neurological: memory loss
Psychiatric: negative

Physical Exam:

Vitals: BP: 150/86 Temp: 98.3 F   Pulse: 86/min, regular RR: 16/min
General: AAO☐3, normocephalic, and atraumatic.

HEENT: PERRLA, EOMI, no papilledema, TM intact, and no nasal discharge.
Neck: supple, no JVD, no lymphadenopathy, and no thyromegaly.
CVS: S1 and S2, regular, no murmurs, no rubs, and no gallops.
Chest: CTA (B/L), no wheezing, no rales, no ronchi, and no scoliosis.
Abdomen: soft, non-tender, non-distended, +BS, and no organomegaly.
Extremities: no clubbing, no cyanosis, and no edema.
CNS: CN 2 – 12 intact, mini-mental score 20/30.
Skin: no rashes
Rectal: palpable, rubbery, and enlarged prostate.

Labs:

CBC, CMP, Lipid panel, PSA, TSH, Vitamin B12 level, U/A, Urine culture, CXR, U/S of prostate, colonoscopy, bone scan, and CT of head.

Results:

PSA 6.O
Other laboratory results came back negative for any acute disease process.

Assessment:

1. Alzheimer's dementia: Patient's daughter complains of long term memory loss which the patient himself denies. CT scan of head negative for any acute process.
2. Urinary Tract Infection: Patient complains of increased frequency. UA is negative for any infection.
3. Benign Prostatic Hypertrophy: Patient complaining of increased frequency and dribbling. Rectal exam positive for prostate enlargement. PSA is slightly elevated.
4. Prostatic cancer: Patient is elderly man with urinary complaints. PSA is slightly elevated. Digital rectal, as well as the transrectal ultrasound shows Prostatic enlargement. Biopsy is negative for any malignancy.

Plan:

1. Health maintenance exams: CBC, CMP, lipid profile, colonoscopy, bone scan.
2. UA for urinary complaints.
3. Start Tamsulosin for both Benign Prostatic Hypertrophy and HTN.
4. Start statins for Dyslipidemia.
5. Start Donepezil to help slow Alzheimer's dementia.
6. Start Alendronate, Calcium & Vitamin D supplements to slow progression of Osteoporosis.
7. Follow up in 1 month.

CASE STUDY 2: Asthma

Chief Complaint: Patient complains of an asthma attack with chest pain, shortness of breath, chest tightness, rushing heart beats, difficulty walking or standing.

History of Present Illness:

Patient is a 39 year old African American female who presents to the Emergency Room with chest pain, shortness of breath, chest tightness, tachycardia and palpitations. Standing up made the symptoms worse while sitting down made it better. Humidity and smoke made the symptoms worse. The patient has a history of asthma for the past 25 years and has had asthma attacks before but none was as intense as this nor has she ever been intubated for her asthma in the past. She denies any fever, wheezing, coughing. She is on Albuterol and home nebulizer treatment but she has not used them for the past 1-2 months. She admits to feeling tired lately. She is distressed and worried because of her financial situation.

Past Medical History: Diabetes, Asthma for approximately 25 years, Hypertension, Angiogram 5 years ago.
Past Surgical History: Angiogram 5 years ago.
Family History: DMII, HTN.
Allergies: NKDA
Medication: Metformin, Metoprolol, Enalapril, Albuterol, nebulizer.
Social History: no smoking, occasional ETOH consumption, no drug abuse, and married.
OB/GYN: G3 P2

Review of Systems:

General: Feels fatigued at home and has decreased ability to perform daily work; decreased sleep; gained 26 pounds and attributed it to decreased physical activity.
Skin: Negative
HEENT: Rare blurring of vision.
Neck: Negative
Breasts: Negative
Chest: Shortness of breath.
CVS: Heart beating fast, chest pain.
Gastrointestinal: Negative
Urinary: Negative
Musculoskeletal: Occasional body aches.
Neurological: Negative
Psychiatric: Nervousness

Physical Exam:

Vitals: BP: 150/92 Temp: 98.0 F Pulse: 110/min RR: 23/min O2 Sat 94% on 2 L NC
General: AAO3, normocephalic, and atraumatic.
HEENT: PERRLA, EOMI, no papilledema, TM intact, and no nasal discharge.
Neck: Supple, no JVD, no lymphadenopathy, and no thyromegaly.
CVS: S1 and S2, regular, no murmurs, no rubs, and no gallops.
Chest: Coarse breath sounds (B/L), wheezing heard (B/L), rales (B/L).
Abdomen: Soft, non-tender, non-distended, +BS, and no HSM.
Extremities: no clubbing, no cyanosis, no edema.
CNS: CN 2 –12 intact.
Skin: no rashes

Labs:

CBC, CMP, cardiac enzymes, ABG, D-dimer, CXR, and EKG V/Q SCAN.

Results: labs came back within normal limits.

1. ABG PH, 7.42, PCO2 = 38, PO2 = 87, HCO3 = 22
2. EKG – showed normal rate and rhythm.
3. V/Q SCAN – negative for any PE.
4. CXR – no acute findings.

Assessment:

1. Acute Asthma attack: Patient has shortness of breath. There is also a history of Asthma. Patient also has wheezing and rales heard on physical exam. Patient has responded to nebulizer treatment
2. Myocardial Infarction: Patient complains of chest pain. EKG shows no ST elevation or any other abnormality. CXR negative for any acute process. Cardiac enzymes also not elevated.
3. Pulmonary Embolism: Patient complains of shortness of breath and pleuritic chest pain. CXR and EKG negative for any acute process. ABG shows normal lab values. V/Q scan shows now abnormality.

Plan:

1) Start O2 therapy, aspirin, morphine and nitrates.
2) Once MI is ruled out, discontinue the aspirin, morphine and nitrates.
3) Start steroids and nebulizer treatment and assess patient after treatment.
4) Once stabilized, discharge patient on Albuterol, home nebulizer treatment and prednisone for approximately 10 days.

CASE STUDY 3: Congestive Heart Failure

Chief complaint: Worsening of shortness of breath.

History of Present Illness:

This is a 78 year-old male came to the office with a 6 month history of SOB that is aggravated by exertion. He has found that he has to get up at night and open the window to get air. He also complained of bouts of breathlessness accompanied by anxiety. He often needs extra pillows to reduce the number and severity of attacks. There has not been any significant weight gain, nor has there been any swelling of his ankles and legs. Pt also complains of some mild non-productive cough. He denies any chest pain, palpitations, or any other symptom.

Past Medical History: MI 8 years ago, HTN, and DMII.
Past Surgical History: Hernia repair 20 years ago and cardiac catheterization 8 years ago.
Family History: HTN, DMII in both father and mother.
Medications: HCTZ, Metoprolol, Metformin, Glipizide, Aspirin, and Plavix.
Allergies: NKDA
Social History: no smoking, no ETOH, no drug abuse, wife passed away, and lives alone.

Review of Systems:

Skin: negative
HEENT: negative
Neck: negative
Chest: complaining of non productive cough and SOB which is worsening.
Gastrointestinal: negative
Urinary: negative
Musculoskeletal: negative
Neurological: negative
Psychiatric: negative

Physical Exam:

Vitals: BP 140/90 P 84 TEMP 98.0 F, RR 28 O2 Sat 92% on 3L NC
General: AAO☐3, normocephalic, and atraumatic.
HEENT: PERRLA, EOMI, no papilledema, and TM intact.
Neck: Supple, mild JVD, and no lymphadenopathy.
CVS: S1 and S2, regular, no murmurs, no rubs, and no gallops.
Chest: Rales B/L
Abdomen: Soft, non-tender, non-distended, +BS, and no HSM.

Extremities: no clubbing, no cyanosis, and 1+ pitting edema.
NEURO: CN 2-12 intact and no gross neurological deficit seen.
Skin: no rashes

Labs:

CBC, CMP, BNP, ABG, CXR, and 2D ECHO.

Results:

Patient's BNP came back 700, CXR showed some mild pulmonary infiltrates. ECHO showed decreased EF 40% consistent with congestive failure.

Assessment:

1) Congestive heart failure: this was diagnosed by the patient presentation and the echo results and the BNP value.
2) Pulmonary edema: there was mild pulmonary infiltrate but no edema.
3) Pneumonia: this was ruled out as CXR did not show any signs consistent with pneumonia.

Plan:

1) Patient was admitted to the hospital.
2) IV lasix 40 mg.
3) Enalapril 40 mg.
4) Metoprolol 80 mg.
5) Cardiology consulted.

CASE STUDY 4: Generalized Fatigue

Chief Complaint: Fatigue, tiredness and difficulty sleeping.

History of Present Illness:

Patient is a 38 year old Caucasian female who presents to her family physician with decreased energy, decreased appetite, and tiredness for the past 3 months. She was separated from her husband few months ago. She lives in a stressful household with 3 kids and recent financial troubles. She has decreased sleep, about 3-4 hours a night. She also has diarrhea for the past month. Diarrhea is watery, 3-4 times a day, denies seeing any blood. There is occasional abdominal pain. Last diarrhea was the morning before she presented to the clinic. She is worried about developing something aggressive like colon cancer.

Past Medical History: Hypertension
Past Surgical History: appendectomy 15 years ago.
Family History: father has HTN mother has DMII and Hypothyroidism.
Medications: HCTZ, Benadryl occasionally for sleep.
Allergies: NKDA
Social History: no smoking, occasional ETOH, no drugs, separated, lives with her 3 kids, and she is a receptionist
OB/GYN: G3 P3, LMP 1 week ago.

Review of Systems:

General: low on energy, 3-4 hours sleep, and decreased appetite.
Skin: negative
HEENT: negative
Neck: negative
Breasts: negative
Chest: negative
CVS: negative
Gastrointestinal: diarrhea for approximately 1 month.
Urinary: negative
Musculoskeletal: negative
Neurological: negative

Physical Exam:

Vitals: BP: 100/70 Temp: 97.0 F Pulse: 92/min, regular RR: 16/min
General: AAO□3, normocephalic. and atraumatic.
HEENT: PERRLA, EOMI, no papilledema, TM intact, and slightly pale conjunctiva.
Neck: supple, no JVD, no lymphadenopathy, and no thyromegaly.

CVS: S1 and S2, regular, no murmurs, no rubs, and no gallops.
Chest: CTA (B/L), no wheezing, no rales, no ronchi, and no scoliosis.
Abdomen: soft, non-distended, +BS, mild diffuse tenderness, and no organomegaly.
Extremities: no clubbing, no cyanosis, and no edema.
CNS: CN 2 – 12 intact.
Skin: no rashes and mild paleness.
Rectal: negative FOBT.
Speech: speech is fluent with slow rate, rhythm and monotone.
Mood: patient appears worried, sad and depressed.
Perceptions: patient is worried about what will happen in life and how she will be able to cope with her bills/financial troubles. She denies hallucinations in any other sensorium.
Thought Process: Thoughts are clear, logical and coherent, denies delusions or ideas of reference; denies any suicidal or homicidal intentions.

Labs:

CBC.CMP, thyroid panel, lipid panel, urine culture, and stool culture.

Assessment:

1) Decreased appetite, fatigue, and tiredness: Possibly due to malnutrition and dehydration. Patient also admits to diarrhea for approximately one month.
2) Depression: Patient is recently separated with multiple social issues. Patient admits to feeling sad however she denies any suicidal tendencies.
3) Diarrhea: Possibly due to her depression or irritable bowel syndrome.
4) Hypothyroidism: Can cause a depressed mood. Also can causes weight gain but patient admits to weight loss possibly due to her depression.

Plan:

1) IV normal saline.
2) Start metronidazole and discontinue when stool cultures are negative.
3) Attain TSH values.
4) If all labs normal, her symptoms are possible due to depression which is leading to irritable bowel syndrome.
5) Discharge with laxatives.
6) Discharge on Alprozolam and referral to a psychiatrist. Also give diet counseling with increase fiber in her diet.

CASE STUDY 5: Gout

Chief Complaint: Right wrist pain.

History of Present Illness:

This is a 47 year old African American female who presents to the Internal medicine wards for pain in her right wrist and hand for approximately 2 days. The right wrist is swollen and red. The pain is described as a sharp stinging sensation that is constant. The intensity of the pain is described as approximately 8/10. It is made worse my movement of her hand. She denies anything that makes the pain get better. Patient also admits to occasional headaches. Patient denies any nausea, vomiting, shortness of breath, chest pain or abdominal pain.

Past Medical History: HTN, DMII, gout, history of carpal tunnel syndrome.
Past Surgical History: cholesystectomy
Family History: father and mother both have HTN and DMII.
Medications: Metoprolol, HCTZ, Metformin, Glipizide, Aspirin, and Allopurinol.
Allergies: NKDA
Social history: no smoking, ETOH occasionally, and no recreational drugs.
Sexual history: sexually active with her husband.

Review of Systems:

General: pain in right wrist and hand.
Skin: right wrist is swollen and red; skin ulcer in right lower leg is present.
HEENT: negative
Neck: negative
Breasts: negative
Gastrointestinal: negative
Chest: negative
Urinary: negative
Musculoskeletal: constant right wrist and hand pain.
Neurological: occasional headaches.
Psychiatric: negative

Physical Exam:

Vitals: BP: 128/88 Temp: 98.0 F Pulse: 89/min, regular RR: 18/min
General: AAO☐3, normocephalic, atraumatic, and obese.
HEENT: PERRLA, EOMI, no papilledema, TM intact, and no nasal discharge.
Neck: supple, no JVD, no lymphadenopathy, and no thyromegaly.
CVS: S1/S2, regular, no murmurs, no rubs, and no gallops.
Chest: CTA (B/L), no wheezing, no rales, no ronchi.

Abdomen: soft, non-tender, non-distended, +BS, and no organomegaly.
Extremities: no clubbing, no cyanosis, no edema, right wrist is tender, swollen and erythematous. There is limited ROM due to pain. Pain is localized to whole wrist and not localized to specific digits of the hand. Patient has a grade 2 ulcer in right medial ankle. There is no discharge and it is non tender.
CNS: CN 2-12 intact and no gross neurological deficit seen.
Skin: no rashes and grade 2 ulcer in medial right ankle.

Labs:

CBC, CMP, ANA, RF, ESR, HGBA1C, uric acid level, XRAY of the R hand, joint aspiration, and wound culture.

Results:

1. Uric Acid (Serum): 12 mg/dL
2. Uric Acid (Urine): 1000 mg/day.
3. Joint Aspiration: Needle-like negatively birefringent crystals.
4. X-Ray of Wrist: Negative for any fractures; no signs of osteoarthritis.
5. Wound Cultures of Leg Ulcers: Negative of bacterial growth.

Assessment:

1) Gout: Patient has history of gout. Wrist is tender and inflamed. Joint aspiration came back positive for negatively birefringent crystals that were shaped like needles. Uric acid was levels were high in both the serum and urine.

Plan:

1) Discontinue Allopurinol and Aspirin.
2) Start Colchicine and NSAID'S for pain.
3) Once pain subsides and acute process has resolved, restart Allopurinol.
4) Wound debridement.
5) Maintain glucose levels with insulin.

CASE STUDY 6: *Hematochezia*

Chief Complaint: Hematochezia; bleeding per rectum.

History of Present Illness:

This is a 57 year old African American male who presented to the ER complaining of bloody stools. The patient said that the blood coats the stool and is bright red color which has been going on for about 1 month. But over the past 3 days it has gotten worse that's when he decided to come to the ER. Pt also complains of occasional abdominal pain. Pt denies any nausea, vomiting, fever or chills, denies any cramping or straining with bowel movements or SOB, or palpitations. Pt also complains of some nonproductive cough. Pt is a known HIV+ and chronic renal failure on dialysis.

Past Medical History: HTN, HIV+ (1990), end stage renal disease.
Past Surgical History: midline surgical scar does not know why he had the surgery.
Family History: father has HTN, DMII and diverticulitis, mother HTN.
Medications: Propranolol, HCTZ, Famotidine, on dialysis 3/week.
Allergies: NKDA
Social History: Smoker with 10 packs year history, ETOH every weekend, and cocaine use occasionally.

Review of Systems:

Skin: negative
HEENT: negative
Neck: negative
Chest: complaining of non productive cough.
Gastrointestinal: bright red blood per rectum filling in the toilet bowl.
Urinary: negative
Musculoskeletal: negative
Neurological: negative
Psychiatric: negative

Physical Exam:

Vitals: BP: 142/90, P 80, TEMP 98.0 F RR 20 O2 Sat 95% on 2 L NC
General: AAO☐3, normocephalic, and atraumatic.
HEENT: PERRLA, EOMI, no papilledema, and TM intact.
Neck: supple, no JVD, no lymphadenopathy, and no thyromegaly.
CVS: S1/S2, regular, no murmurs, no rubs, and no gallops.
Chest: CTA (B/L), no wheezing, no rales, and no ronchi.
Abdomen: soft, non-tender, non-distended, +BS, and no HSM.
Extremities: no clubbing, no cyanosis, and no edema.

CNS: CN 2-12 intact and no gross neurological deficit seen.
Skin: no rashes and small macules on the lower B/L extremity.
Rectal: positive for fresh blood and no masses felt.

Labs:

CBC, CMP, Gram stain for sputum, CXR, and CD4 count viral load.

Results:

Patients' H and H came back 7.4 and 28.0 other labs were normal, CD 4 VIRAL load was sent out.

Assessment:

1) Hemorrhoids- rectal exam helped to rule that out and the patient had no history of it.
2) Anal fissures- on examination there were no signs of fissures or abscess.
3) Diverticulosis- A GI consult was ordered for a colonoscopy after bleeding stops.
4) Angiodysplasia- waiting for colonoscopy.
5) CMV colitis- waiting for colonoscopy.

Plan:

1) Patient was transfused 2 units of blood based on his HGB 7.4 and clinical picture.
2) Azithromycin 500mg q daily for PCP prophylaxis.
3) Trimethoprim-Sulfamethoxazole prophylaxis for toxoplasmosis.
4) GI consult for possible colonoscopy after bleeding stops.

CASE STUDY 7: *Myocardial Infarction*

Chief Complaint: Chest pain that moved down to arms.

History of Present Illness:

Patient is a 58 year old male who presented to the ER with chest pain that radiated to his arms. He described a sharp chest pain that started at his chest and later moved down to both of his arms. Onset was one night prior to presenting to the ER. The pain lasted thru the entire night. Each occurrence lasted about 5 minutes and was relieved by sitting down. After the patient went back to sleep he kept waking up about every hour due to the pain. He recalls having more than 10 of such instances. Any extensive activity would aggravate the pain. The pain was not getting worse and it was the same intensity throughout. The patient also noticed shortness of breath. There were no palpitations, headaches, nausea, vomiting or diaphoresis.

Past Medical History: Hepatitis B, Diabetes, and DL.
Past Surgical History: none
Family History: father and mother both had HTN, DMII, and CAD.
Medications: Aspirin, Glipizide, Metformin, and Statins.
Allergies: NKDA
Social History: 20 pack year history, occasional ETOH, no recreational drugs, and retired mail man

Review of Systems:

General: feels a little low of energy for the past three months.
Skin: negative
HEENT: negative
Neck: negative
Chest: SOB and chest pain.
CVS: negative
Gastrointestinal: negative
Urinary: negative
Musculoskeletal: negative
Neurological: negative
Psychiatric: nervousness and depression.

Physical Exam:

Vitals: BP 140/92,   P95,   RR20, O2 Sat 95% ON RA
General: AAO☐3, normocephalic and atraumatic, obese.
HEENT: PERRLA, EOMI, no papilledema, TM intact, and no nasal discharge.
Neck: supple, no JVD, no lymphadenopathy, and no thyromegaly.

CVS: S1 and S2, regular, no murmurs, no rubs, and no gallops.
Chest: CTA (B/L), no wheezing, no rales, and no ronchi.
Abdomen: soft, non-tender, non-distended, +BS, and no organomegaly.
Extremities: no clubbing, no cyanosis, and no edema.
CNS: CN 2 – 12 intact
Skin: no rashes

Labs:

CBC, CMP, cardiac enzymes, ABG, CXR, and EKG.

Results:

1. Cardiac enzymes came back within normal limits.
2. EKG showed normal sinus rhythm with no ST segment elevations.
3. CXR did not show any sign of infiltration.

Assessment:

1. Myocardial Infarction: Patient complains of chest pain and has increased risk factors. EKG shows no ST elevation or any other abnormality. CXR negative for any acute process. Cardiac enzymes also not elevated.
2. GERD: Patient complains of chest pain that is constant and occurs at night. Patient has been ruled out for an MI. Patient has responded to the GI cocktail.

Plan:

1. Start O2 therapy, aspirin, morphine and nitrates.
2. Once MI is ruled out, discontinue the aspirin, morphine and nitrates.
3. Administer a GI cocktail, Maalox and diphenhydramine.
4. Discharge after attaining negative cardiac enzymes at 3 different time intervals.
5. Discharge on PPI.

CASE STUDY 8: *Pneumonia*

Chief Complaint: Cough and shortness of breath.

History of Present Illness:

27 year old male comes to the clinic complaining of cough and shortness of breath for 2 days. Patient says that his cough is productive and has yellow colored sputum. Patient also complains of fever of about 102 F and body aches for the past couple of days. Patient denies any chills, night sweats, nausea, vomiting, or recent travel out of the country. Patient said he tried OTC medicine acetaminophen for his fever but says he still feels the same.

Past Medical History: Asthma
Past Surgical History: Appendectomy (1995)
Family History: non contributory
Medications: Albuterol inhaler PRN
Allergies: NKDA
Social History: truck driver, smokes 1PPD for the past 10 years, + ETOH once/ wk, cannabis use once/wk, and occasional IV drug abuse.
Sexual History: sexually active with multiple females, no use of condoms and occasional sex with prostitutes.

Review of Systems:

Skin: no complaints
HEENT: negative
Neck: negative
Chest: cough productive yellow sputum and no chest pain.
Gastrointestinal: negative
Urinary: negative
Musculoskeletal: WNL
Neurological: occasional headaches.
Psychiatric: negative

Physical Exam:

Vitals: BP 125/82 HR 89 Temp103 F RR 20 O2 Sat 92% on 2L NC.
General: alert and oriented X3, not in any apparent distress.
HEENT: a traumatic, PERRLA, EOM intact, and throat moist mucosa.
Neck: supple and + cervical lymphadenopathy.
Lungs: crackles at the bases and coarse breath sounds B/L, and no wheezes.
CVS: S1 and S2 audible, RRR, no murmurs, and no gallops.
Abdomen: soft, non tender, no distension, +BS, and no HSM.

Extremities: no edema, no cyanosis, and no rashes.
Neuro: CN 2-12 intact and no gross neurological deficits.

Labs:

CBC, CMP Sputum gram stain, and HIV testing.

Results:

CXR showed infiltrates in left lower lobe consistent with lobar pneumonia.

Assessment:

1. Pneumonia : CXR confirmed the diagnosis sputum gram stain also came back positive for streptococcus pneumonia.
2. Tuberculosis: CXR ruled that out.
3. HIV: HIV testing the ELISA came back negative.
4. Lung abscess: CXR did not show any signs of abscess.
5. Asthma: Started on Albuterol nebulizer.

Plan:

1. Patient is started on Levaquin 750 mg q day IV.
2. Patient was also started on IV fluids 0.9% NS.
3. Ibuprofen 200 mg q 4-6 hours for fever.
4. Albuterol nebs q 6 hours.

CASE STUDY 9: Rhabdomyolysis

Chief Complaint: Chest pain.

History of Present Illness:

This is a 27 year old male who presented to us via the emergency room complaining of chest pain for the past 24 hours. Pt said that the pain radiated to his left arm and shoulder and graded the pain 7 out of 10 on the pain scale. Pt said that the pain was stabbing in quality and was not alleviated by any OTC meds. Pt also mentioned that the pain started after vigorous exercise a night before which involved lifting heavy weights and running for couple of miles. Pt also complained of some SOB and said that his body felt sore. Denies any palpitations, nausea or vomiting.

Past Medical History: wrist fracture (2003).
Past Surgical History: hernia repair
Family History: father alive and has HTN, DM II, mother alive has Hyperthyroidism, HTN, has one brother alive and well.
Medications: none
Allergies: NKDA
Social History: no smoking, ETOH occasionally, and no recreational drugs.
Occupation: personal athletic trainer.
Sexual history: sexually active with his girlfriend.

Review of Systems:

Skin: no complaints
HEENT: negative
Neck: negative
Chest: substernal pain radiating to left arm and shortness of breath.
Gastrointestinal: negative
Urinary: negative
Musculoskeletal: calf pain and soreness all over the body.
Neurological: negative
Psychiatric: negative

Physical Exam:

Vitals: BP 125/82 HR 89 TEMP 99.0 F RR 20 O2 Sat 98% on RA
General: Alert and oriented X3, not in any apparent distress.
HEENT: atraumatic, PERRLA, EOM intact, and throat moist mucosa.
Neck: supple and no lymphadenopathy
Lungs: CTA B/L
CVS: S1 and S2 audible, RRR, no murmurs, and no gallops.

Abdomen: soft, non tender, non distension, +BS, and no HSM.
Extremities: no edema, no cyanosis, and no rashes.
Neuro: CN 2-12 intact and no gross neurological deficits.

Labs:

CBC, CMP serum CK, cardiac enzymes CKMB, Troponin, LDH, D-dimer, UA, CXR, EKG, and 2D-ECHO.

Results:

1. Labs came back within normal limits except for a serum CK of 5000 and BUN/CR 6.00.
2. CXR was consistent with normal heart and lungs, EKG did not show any signs of ischemia.
3. 2D ECHO showed EF of 55% (mainly within normal limits) no other gross abnormalities.

Assessment:

1) Rhabdomyolysis: after ruling out other pathological abnormalities serum CK was 5000.
2) MI: cardiac enzymes ruled that out which came back negative.
3) Pericarditis: pt clinical symptoms were not that consistent with this plus the echo did not show any sign suggestive of pericarditis.
4) Pulmonary embolism: D- dimmers were negative
5) GERD: unlikely

Plan:

1) Patient was started on IV fluid 5% dextrose solution.
2) Patient was given acetaminophen for his pain q6.
3) Cardiology consult was ordered to rule out any other cardiac related abnormality.

CASE STUDY 10: Transient Ischemic Attack

Chief Complaint: Blurred vision, difficulty speaking, and brief loss of consciousness.

History of Present Illness:

This is an 89 year old Caucasian male who presents to the ER complaining of blurred vision, cloudiness, confusion, inability to understand and inability to speak while talking to his daughter on the phone on Friday night. She called the emergency room and the patient was picked up from his house in an unconscious state. He regained consciousness immediately when he reached ER. He experienced no abdominal pain, no polyuria, polydipsia, or polyphagia. He noticed no chest pain, no nausea or vomiting, no tachycardia and no kussmaul's respirations. This is the first time he experienced these symptoms. He denies any illicit drug abuse. This incident has left him worried about future occurrences that hinder his ability to keep up with daily chores.

Past Medical History: HTN, DMII, pacemaker.
Past Surgical History: Cataract removed about 10 years ago, Hernia repair (long back).
Family History: HTN, DMII mother and father passed away due to unknown cause.
Medications: Lisinopril, Plavix, Furosemide, PPI, Finasteride, MVT, Allopurinol, and Metoprolol,
Allergies: NKDA
Social History: no smoking, no ETOH, no recreational drugs, and wife passed away about 10 years ago.

Review of systems:

Skin negative
HEENT: blurry vision and congested nose.
Neck: negative
Chest: negative
CVS: negative
Gastrointestinal: negative
Urinary: negative
Musculoskeletal: negative
Neurological: negative
Psychiatric: negative

Physical Exam:

Vitals: BP: 140/87 Temp: 98.8 F Pulse: 90/min, regular RR: 18/min
General: AAO☐2, normocephalic, and atraumatic.
HEENT: PERRLA, EOMI, no papilledema, TM intact, and no nasal discharge.
Neck: supple, no JVD, no lymphadenopathy, and no thyromegaly.

CVS: S1 and S2, regular.
Chest: CTA (B/L), no wheezing, no rales, and no ronchi.
Abdomen: soft, non-tender, non-distended, +BS, and no organomegaly.
Extremities: no clubbing, no cyanosis, and no edema.
CNS: CN 2-12 intact, Mini Mental 22/30.
Skin: no rashes

Labs:

CBC, CMP, ACCUCHECK, cardiac enzymes, CXR, EKG, and CT of head.

Results:

1. STAT Glucose: 75 mg/dl.
2. EKG: Unable to interpret acutely due to pacemaker but within normal limits.
3. CXR: Negative for any acute process.
4. Troponin T: 0.12 µg/L.
5. CK-MB: 14.0 U/L.
6. CT of Head: Negative for Ischemic or Hemorrhagic process.

Assessment:

1) Transient Ischemic Attack: Patient has multiple risk factors. Patient also had an episode of loss of consciousness. There is also an episode of aphasia and blurry vision. CT scan of head shows no hemorrhagic or ischemic process.
2) Dehydration: Elderly patient who is possibly dehydrated since he lives alone with multiple problems and possibly unable to take care of own self completely.
3) Stroke: Patient has multiple risk factors. Patient also admits to loss of functionality of part of his extremities with aphasia. Patient also admits to blurry vision.
4) Hypoglycemia: glucose is slightly low. Hemoglobin A1C is slightly elevated indicating poorly controlled diabetes. Patient also complains of blurry vision.

Plan:

1) IV normal saline, O_2 via NC.
2) Attain CT scan of Head, Arteriography, Doppler Ultrasound of Carotid arteries.
3) If CT scan normal; possible MRI of head.
4) Get EKG, Echocardiograph.
5) Get a neurological consult.
6) Admit to wards for further observation.

CASE STUDY 11: Gastrointestinal Bleeding

Chief Complaint: 22 years old male presents with complaint of blood in his stools.

History of Present Illness:

He has been noticing blood in his stools for the past one month but decided to come today to see a doctor because the bloody stools did not go away. For the past one month the patient says there has been no change in his bowel movements or the amount or quantity of blood in his stools. He says that the stools are not bloody all the time but occasionally. He describes the stool as brown blood mixed with stool with no frank fresh bleeding. He says that he is a college student and has to study late at night and drinks lot of coffee which is causing him lot of reflux and sometimes causes him epigastric pain. He denies any hematemesis or any hematuria. He denies any fever, no N/V/D/C, no recent travel, no sick contacts, no palpitations, no headaches, no dizziness, no weight loss.

Past Medical History: none

Past Surgical history: none

Family History: HTN

Medications: antacids for reflux.

Allergies: Penicillin

Social History: (+) Smoke No, (+) ETOH, on the weekends doesn't keep account, (+) drugs, marijuana occasionally.

Previous Episode: ongoing on and off for one month.

Hospitalizations: none

Personal History: none

Physical Exam:

Vitals: BP 122/ 78, RR 18, Pulse 81, Temp. 98.3 F, BS 67 mg/dl.

General: Patient is alert and oriented and in no acute distress.

HEENT: normocephalic, atraumatic no pallor, no icterus, non-exudative, non-erythematous, and no signs of thrush.

Neck: supple, no JVD, no lymphadenopathy, and no thyromegaly.

CVS: S1 and S2, RRR, no S3 or S4, no rubs, gallops, and murmurs could be appreciated. (-) JVD, (-) Pedal edema.

Chest: CTA B/L, no wheezes, and no ronchi.

Abdomen: soft, non-tender, scaphoid, tympanic, with normal liver edge.

Extremities: no active lesions or scars, no clubbing, cyanosis, or edema.

CNS: CN 2-12 intact.

Skin: no rashes

Labs:

CBC, CMP, other labs depending on clinical suspicion.

Assessment:

1. GI bleeding could be upper or lower with upper more likely.

Plan:

1. Schedule the patient for an endoscopy or a colonoscopy.

2. Refer to GI specialist.

3. Treat appropriately depending on the etiology.

CASE STUDY 12: **Hematuria**

Chief Complaint: 62 year old male presents with complaint of blood in his urine to the outpatient office.

History of Present Illness:

Per patient it started 2 days ago and this morning. He says that for the past 2 years he has had difficulty urinating. He mostly complains of increased in nocturnal visits to the bathroom to urinate but no flank pain. He also says that occasionally he can't make it to the bathroom on time and passes urine. He also says that he has a sensation of incomplete bladder emptying even when he goes to the bathroom. He says that for the past 1 year his urinary stream has also decreased and reports hesitancy when he has to go. He denies any dysuria or any pain or any pus in his urine or any burning sensation but does report he has to strain a lot to go to the bathroom. This morning he saw blood in his urine and decided to get it checked out by a doctor. He denies any fever, no N/V/D/C, no palpitations, no headaches, no dizziness, no abdominal pain, no weight loss.

Past Medical History: HTN and DMII.

Past Surgical History: appendectomy when he was 14

Family History: HTN, DMII

Medications: Lisinopril, Insulin, Metformin.

Allergies: NKDA

Social History: (+) Smoke 1 pack X 30 years, (+) ETOH 2 beers Qd, (-) Drugs.

Previous Episode: 2 days ago.

Hospitalizations: for a MVA 10 years ago

Physical Exam:

Vitals: BP 142/ 88, RR 18, Pulse 88, Temp. 98 F, BGS 87 mg/dl.

General: Patient is alert and oriented and in no acute distress.

HEENT: normocephalic, atraumatic, no pallor, no icterus, non-exudative, non-erythematous, no signs of thrush, (+) poor dentition.

Neck: supple, no JVD, no lymphadenopathy, and no thyromegaly.

CVS: S1 and S2, RRR, no S3 or S4, no rubs, gallops, and murmurs could be appreciated. (-) JVD, (-) Pedal edema.

Chest: CTA B/L, no wheezes, and no ronchi.

Abdomen: soft, non-tender, scaphoid, tympanic, with normal liver edge, and mild right CVA tenderness.

Extremities: no active lesions or scars, no clubbing, cyanosis, or edema.

CNS: CN 2-12 intact.

Skin: no rashes

Labs:

CBC, CMP, other labs depending on clinical suspicion.

Assessment:

1. Hematuria

Plan:

1. Rule out identifiable causes such as BPH, prostate cancer, urolithiasis, renal cell carcinoma.

2. Treat appropriately depending on the etiology.

CASE STUDY 13: Asthma

Chief Complaint: 52 years old male presents with complaint of shortness of breath and difficulty breathing.

History of Present Illness:

Per patient it started 2 days ago and was progressive in character until he had difficulty taking breaths felt tired, fatigued and decided to come to the hospital ER. There is non-productive cough, but no fever, no N/V/D/C, no palpitations, no headaches, no dizziness, no abdominal pain, no weight loss.

Past Medical History: HTN, DMII, Asthma since 7 years old, GERD.

Past Surgical History: none

Family History: HTN, DMII

Medications: Albuterol, Lisinopril, Insulin, Metformin, and antacids.

Allergies: NKDA

Social History: (+) Smoke 1 pack X 25 years, (+) ETOH, and (-) Drugs.

Previous Episode: the patient was hospitalized 3 times in the past 2 years for asthma exacerbation.

Hospitalizations: many due to various reasons.

Physical Exam:

Vitals: BP 136/ 88, RR 20, Pulse 81, Temp. 97 F, BGS 173 mg/dl.

General: Patient is alert and oriented

HEENT: normocephalic, atraumatic, no pallor, no icterus, non-exudative, non-erythematous, no signs of thrush, and (+) poor dentition.

Neck: supple, no JVD, no lymphadenopathy, and no thyromegaly.

CVS: S1 and S2, RRR, no S3 or S4, no rubs, gallops, and murmurs could be appreciated. (-) JVD and (-) Pedal edema.

Chest: Bilateral wheezes heard at the lower lung bases.

Abdomen: soft, non-tender, scaphoid, tympanic, with normal liver edge.

Extremities: no active lesions or scars, no clubbing, cyanosis, or edema.

CNS: CN 2-12 intact.

Skin: no rashes

Labs:

CBC, CMP, other labs depending on clinical suspicion.

Assessment:

1. Reactive Airway Disease, Asthma

Plan:

1. Admit the patient and give him oxygen and start him on Albuterol Nebulizers and start IV Methyl prednisone.
2. Check a CBC, BMP, ABG's, check his oxygen saturation.
3. CXR, if warranted and start him on Levaquin.
4. Reassess his situation if stable discharge with asthma prevention education and if needed medication changes.

Surgery

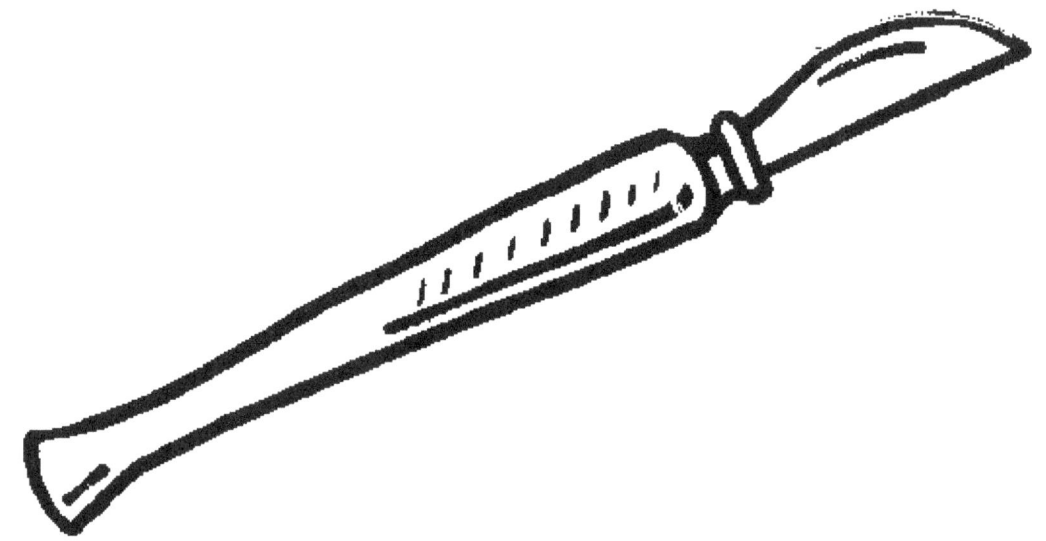

Surgery Case 1: Acute Cholecystitis

Chief Complaint: 42 years old female comes to the ER presenting with the complaint of abdominal pain.

History of Present Illness:

She said the pain is located in the right upper quadrant and is radiating to the shoulder with an intensity of 8/10. She describes the pain as burning in character, constant, and says, "It feels like someone is squeezing me from inside." The pain started two hours ago when she and her family were having lunch at a restaurant. She says the pain has been present for almost 5 months but was never this bad and used to relieve by drinking milk, taking antacids, and PPI. It is sometimes accompanied by nausea and has once vomited yellowish/ non-bloody fluid 2 weeks ago. She denies any diarrhea or constipation or any appetite or weight changes.

Past Medical History: DM II and HTN.
Past Surgical History: Appendectomy at age 13
Allergies: PCN, latex.
Medications: Metformin, Insulin, Captopril, antacids.
Hospitalizations: at age 13 for appendectomy
Social History: Smoke: NO   Illicit Drugs: NO ETOH: yes, socially.
Family History: father died of MI at age 63, mom died of colon cancer 58

Physical Exam:

Vitals: BP 128/ 69   Temp 99.7 F   P 87   RR 19

General: Patient is alert and oriented and in significant pain.
HEENT: PERRLA, moist mucus membranes, and NC/AT.
Lungs: CTA B/L, no ronchi, no wheezes, and good effort.
Cardiac: regular rate and rhythm without murmurs, gallops or rubs.
Abdomen: soft, tender in the right upper quadrant, non-distend, no HSM, patient exhibits positive Murphy's sign.
Extremities: without clubbing, cyanosis or lower edema or calf tenderness.
Skin: warm, dry, no rash, no cellulitis, no jaundice.
Neuro: A focal

Assessment:

1. Possibility of Acute Cholecystitis

Plan:

1. Do a blood test check AST, ALT, Bilirubin, ALK Phos, amylase, lipase.

2. Ultrasound upper abdomen, possibly do a HIDA scan.

3. If above test indicate Acute Cholecystitis send the patient for surgery.

Surgery Case 2: Acute Appendicitis

Chief Complaint: 16 years old male comes to the ER presenting with the complaint of abdominal pain.

History of Present Illness:

Patient said the pain is located in the right lower quadrant and is radiating to the umbilicus with an intensity of 9/10. He describes the pain as burning/ pinching in character, constant, and says, "It is one of the worst pains he has ever felt." The pain started two hours ago when he and his friends were skateboarding at the park. It is accompanied by fever, nausea, and vomiting. He denies any diarrhea or constipation or any weight changes. But does say he doesn't feel like eating or drinking anything.

Past Medical History: Asthma
Past Surgical History: tonsillectomy at age 4
Allergies: seasonal allergies-dust, pollen, etc
Medications: Albuterol and Loratadine.
Hospitalizations: at age 4 for tonsillectomy
Social History: Smoke: NO   Illicit Drugs: Yes, Marijuana occasionally   ETOH: No
Family History: Mom and Dad are alive and healthy

Physical Exam:

Vitals: BP 124/ 66   Temp 101.7 F   P 76   RR 17

General: Patient is alert and oriented and in significant pain.
HEENT: PERRLA, moist mucus membranes, and NC/AT.
Lungs: CTA B/L, no ronchi, no wheezes, and good effort.
Cardiac: regular rate and rhythm without murmurs, gallops or rubs.
Abdomen: soft, tender in the right upper quadrant, non-distend, no HSM, patient exhibits positive Psoas, Rowsing's, Obturator signs.
Extremities: without clubbing, cyanosis or lower edema or calf tenderness.
Skin: warm, dry, no rash, no cellulitis, and no jaundice.
Neuro: A focal

Assessment:

1.   Possibility of Acute Appendicitis

Plan:

1. Do a blood test check AST, ALT, Bilirubin, ALK Phos, amylase, lipase, WBC count, and CBC, maybe ultrasound or a CT but physical exam is highly suggestive.

2. If above test indicate Acute Appendicitis send the patient for surgery.

Surgery Case 3: Inguinal Hernia

Chief Complaint: 76 years old female comes to the clinic presenting with the complaint of blood in her stools and painful protruding inguinal mass.

History of Present Illness:

She said she has had the inguinal mass for a couple of years but has never caused her any discomfort or any pain but for the past 7 days she has felt an increased pressure and pain sensation in the area. She describes the pain as non-radiating, throbbing in character, and constant. She has had one bloody bowel movement today mixed with mucus so she decided to check herself in the hospital. She has a history of chronic constipation for 1 year consisting of bowel movements of twice a week. She has no sense of urgency in her bowel movements. She denies any history of fever, chills, nausea, vomiting, recent history of travel, or any sick contacts. She has had no weight or appetite changes. Her diet consists of lot of unhealthy food devoid of fiber.

Past Medical History: DM II, HTN, DL.
Past Surgical History: cholecystectomy at age 43
Allergies: NKDA
Medications: Metformin, Glipizide, Statins, HCTZ.
Hospitalizations: at age 43 for cholecystectomy
Social History: Smoke: yes, 1 pack qd x 60 years   Illicit Drugs: NO   ETOH: No
Family History: unremarkable

Physical Exam:

Vitals: BP 110/ 72   Temp 98.5 F   P 72   RR 16

General: Patient is alert and oriented to time, place, and person. Well groomed, seems in agonizing pain in the inguinal area.
HEENT: PERRLA, moist mucus membranes, and NC/AT.
Lungs: CTA B/L, no ronchi, no wheezes, and good effort.
Cardiac: regular rate and rhythm without murmurs, gallops or rubs.
Abdomen: soft, tender, non-distended, no HSM, non-reducible swelling in the left groin with no other palpable masses.
Rectal Exam: FOBT negative and no abnormalities.
Extremities: without clubbing, cyanosis or lower edema or calf tenderness.
Skin: warm, dry, no rash, no cellulitis, and no jaundice.
Neuro: A focal

Assessment:

Physical exam consistent with incarcerated inguinal hernia

Plan:

1. Schedule for an emergency hernia repair.

Surgery Case 4: Hyperthyroidism

Chief Complaint: 28 years old female comes to the clinic presenting with the complaint of dysphagia and odynophagia.

History of Present Illness:

She said since two months she has been noticing more difficulty swallowing solid food only no liquids. During this time she has also noted that she feels hot more than usual and has started to notice a hand tremor that occurs while at rest as well as when she is doing other activity. For the past one month she has also had watery diarrhea with no mucus that is non-painful in nature. She has also had difficulty sleeping in the night and reports symptoms of insomnia. She also reports that during this time her menstrual period intervals has changed as they are more frequent but she says it might be due to the stress of her new job. She denies any history of fever, chills, nausea, vomiting, abdominal pain, recent history of travel, or any sick contacts. She has had no weight or appetite changes. Her diet consists of lot of green vegetables as she is a vegetarian.

Past Medical History: unremarkable
Past Surgical History: unremarkable
Allergies: latex
Medications: Acetaminophen occasionally for pain
Hospitalizations: none reported
Social History: Smoke: yes, 1 pack qd x 12 years   Illicit Drugs: NO   ETOH: yes, occasionally
Family History: Mom passed away car accident, father is alive and a veteran.

Physical Exam:

Vitals: BP 109/ 78   Temp 100.6 F P 102   RR 19
General: Patient is alert and oriented to time, place, and person. Well groomed, seems nervous and overly anxious about her symptoms.
HEENT: PERRLA, moist mucus membranes, NC/AT, lid lag, and exophthalmus.
Lungs: CTA B/L, no ronchi, no wheezes, and good effort.
Cardiac: regular rate and rhythm without murmurs, gallops or rubs.
Abdomen: soft, non-tender, non-distended, and no HSM.
Extremities: without clubbing, cyanosis or lower edema or calf tenderness, and fine tremor.
Skin: warm, moist, no rash, no cellulitis, and no jaundice.
Neuro: A focal

Assessment:

Physical exam consistent with hyperthyroidism (exophthalmus)

Plan:

1. Do a blood test check TSH, T4, T3, WBC count, and CBC.

2. Check FSH, LH, Prolactin levels.

3. If above test indicate Hyperthyroidism discuss treatment options (methimazole, propylthiouracil), radioactive iodine ablation, but eventually discuss the benefits of thyroidectomy.

Surgery Case 5: Hemorrhoids

Chief Complaint: 23 years old female comes to the clinic presenting with the complaint of blood in her stools and pain during defecation.

History of Present Illness:

She said since one month she has been noticing blood mixed with her stools and pain during every bowel movement. During the last the last 5 days she has had tenesmus and bloody diarrhea that is painful in nature after she started taking laxatives for the constipation. She has a history of chronic constipation for 1 year consisting of bowel movements of twice a week. She has no mucus or a sense of urgency in her bowel movements. She denies any history of fever, chills, nausea, vomiting, abdominal pain, recent history of travel, or any sick contacts. She has had no weight or appetite changes. Her diet consists of lot of unhealthy food devoid of fiber.

Past Medical History: DM I
Past Surgical History: unremarkable
Allergies: NKDA
Medications: laxatives and Insulin.
Hospitalizations: at age 17 for MVA
Social History: Smoke: yes, 1 pack qd x 6 years   Illicit Drugs: NO   ETOH: yes, weekends, doesn't keep account
Family History: mom and dad are alive and healthy.

Physical Exam:

Vitals: BP 111/ 69   Temp 98.2 F   P 72   RR 19

General: Patient is alert and oriented to time, place, and person. Well groomed, seems distressed about her symptoms.
HEENT: PERRLA, moist mucus membranes, and NC/AT.
Lungs: CTA B/L, no ronchi, no wheezes, and good effort.
Cardiac: Regular rate and rhythm without murmurs, gallops or rubs.
Abdomen: soft, non-tender, non-distended, and no HSM.
Rectal Exam: signs of prolapsed rectal mucosa and anal trauma.
Extremities: Without clubbing, cyanosis or lower edema or calf tenderness.
Skin: warm, dry, no rash, no cellulitis, and no jaundice.
Neuro: A focal

Assessment:

Physical exam consistent with hemorrhoids

Plan:

1. Do a blood test check AST, ALT, Bilirubin, ALK Phos, amylase, lipase, WBC count, and CBC, FOBT, H pylori testing, maybe endoscopy or colonoscopy to rule out other GI pathologies but physical exam is highly suggestive.

2. If above test indicate Hemorrhoids send the patient for surgery (hemorrhoidectomy).

Surgery Case 6: Renal Cell Carcinoma

Chief Complaint: Hematuria.

History of Present Illness:

This is a 40 year old African American female who presented to the ER complaining of painful hematuria, abdominal pain and right flank pain. Patient said that her lower abdominal pain and flank pain has been there for about 2 weeks for which she went to her regular primary care physician who gave her levaquin and scheduled her for an ultrasound but the pain got so severe she had to come to the hospital. Patient describes the pain as constant cramping pain and graded the pain 7/10 on the pain scale. Patient said that for the past 2 days she also started to see some blood in her urine. Patient also complains of fever, nausea and vomiting, + urgency + frequency – dysuria.

Past Medical History: UTI, HTN.
Past Surgical History: Gun shot wound repair (1980).
Family History: father and mother both have HTN, DMII.
Medications: clonidine, levaquin, alprazolam, clonzepam.
Allergies: NKDA
Social History: no smoking, no ETOH, cannabis occasional use.
Sexual History: sexually active with her boyfriend.

Review of Systems:

General: patient complained of some mild fatigue and weight loss some time.
Skin: no complaints
HEENT: negative
Neck: negative
Chest: negative
Gastrointestinal: complains of pain lower abdomen, nausea, vomiting.
Urinary: + urgency + frequency + hematuria
Musculoskeletal: WNL
Neurological: negative.
Psychiatric: negative

Physical Exam:

Vitals: BP 100/71 P90 TEMP 99.0 F RR 20 O2 Sat 98% on RA
General: Alert and oriented X3.
HEENT: atraumatic PERRLA, EOM intact, and throat moist mucosa.
Neck: supple, no thyromegaly, and no masses.
Lungs: CTA B/L no wheezes, no rales, and no crackles.
CVS: S1 and S2 audible, RRR, no murmurs, and no gallops.

Abdomen: soft, midline scar from previous surgery, + tenderness with palpation in the right lower quadrant, and lower abdomen, no distension +BS, and no HSM.
Extremities: no edema, no cyanosis, and no rashes.
Neuro: CN 2-12 intact and no gross neurological deficits.
Back: + right CVA tenderness.

Labs:

CBC, CMP, U/A, urine culture, B-HCG, KUB of the abdomen, U/S of the kidneys, and CT of abdomen.

Results:

1) Labs came back with a mild elevation of WBC count no left shift.
2) U/A and culture showed heavy E. coli growth.
3) KUB showed right kidney mass and mild right side obstruction.

Assessment:

1) Right kidney mass (possible renal cell carcinoma): A surgical consult was ordered to biopsy the mass in other to know its finding.
2) UTI: pt was continued on the levaquin.
3) Pregnancy: that was ruled out by the negative B-HCG.
4) Pyelonephritis: her clinical labs did not point toward pyelonephritis and the U/S did not show any finding pertaining to a pyelonephritis.

Plan:

1) Patient was continued on IV levaquin 500 mg.
2) Surgical consult was ordered and the biopsy came back positive for renal cell carcinoma.

Surgery Case 7: Appendicitis

Chief Complaint: Abdominal pain

History of Present Illness:

This is a 20 year old female who presented to the ER complaining of abdominal pain for the past 24 hours. Patient said that the pain started all of a sudden in the middle of the night without any warning. Patient complained that the pain started in umbilical region then moved toward right side without any radiation. Patient graded the pain 9 on a pain scale of 0-10. Patient describes the pain as sharp in quality denies anything that makes his pain better or worse. Patient also complained of two episodes of vomiting since the onset of pain, pt also feels nauseated. Patient denies any diarrhea or bloody stool no SOB or palpitations.

Past Medical History: URI 2 weeks ago.
Past Surgical History: none
Family History: mother and father alive and well.
Medications: none
Allergies: Penicillin
Social History: smokes ½ PPD, Beer on the weekends, and weekly cannabis use.
Sexual History: sexually with boy friend and uses condoms.
OB/GYN: Last LMP was 1 weeks ago, periods are regular, and no history of any STD.

Review of systems:

Skin: no complaints
HEENT: negative
Neck: negative
Chest: negative
Gastrointestinal: complains of pain towards the right lower side.
Urinary: negative
Musculoskeletal: WNL
Neurological: negative.
Psychiatric: negative

Physical Exam:

Vitals: BP 130/90 P90 TEMP 102.9 F RR 20 O2 Sat 95% on RA
General: alert and oriented X3 and pt is in moderate distress.
HEENT: atraumatic, PERRLA, EOM intact, and throat moist mucosa.
Neck: supple, no thyromegaly, and no masses
Lungs: CTA B/L no wheezes, no rales, and no crackles.

CVS: S1 and S2 audible, RRR, no murmurs, and no gallops.
Abdomen: soft, + Tenderness with palpation in the right lower quadrant, mild distension +BS, pt was in distress even with mild touch had severe pain, and no HSM.
Extremities: no edema, no cyanosis, and no rashes.
Neuro: CN 2-12 intact and no gross neurological deficits.

Labs:

CBC, CMP, U/S of the abdomen UA/urine culture, B-HCG, and CT abdomen.

Results:

1) The WBC count came back at 25000 with a left shift.

Assessment:

1) Appendicitis: looking at the clinical picture and the symptoms, appendicitis was the most probable but to confirm a surgical consult was ordered.
2) Ectopic pregnancy: B-HCG was negative and the U/S did not show any evidence of a pregnancy.
3) Ovarian cyst: the U/S did not show any evidence of a cyst.
4) Kidney stone: the CT was negative and showed no signs of a stone.

Plan:

1) Patient was started on IV fluids.
2) Patient was kept NPO.
3) Surgical consult was called for immediate surgical intervention.

Pediatrics

CASE STUDY 1: Otitis media

Chief Complaint: Unusual crying and pulling the right ear.

History of Present Illness:

Patient is an 18 month old male infant who has been crying excessively for the last 2 days. He has been very irritable, more than usual. The parents also state that he has been tugging and pulling his right ear excessively for the last couple of days. The parents admit to him having a fever one day ago but it was subsided by giving him children's Acetaminophen. There are no other complaints. Patient has normal bowel movements. Parents also states that he has decreased appetite and is less playful which is kind of unusual for him. Parents deny any vomiting, diarrhea, or skin rashes.

Past Medical History: None
Past Surgical History: None
Family History: Non contributory
Medications: Acetaminophen
Allergies: NKDA
Birth History: Normal vaginal delivery at 38 weeks and no complications.
Immunizations: Up to date

Review of Systems:

General: very irritable.
Skin: negative
HEENT: right ear tugging.
Neck: negative
Chest: negative
CVS: negative
Gastrointestinal: negative
Urinary: negative
Musculoskeletal: negative
Neurological: negative

Physical Exam:

Vitals: BP: 90/60   Temp: 103.4 F   Pulse: 98/min, regular   RR: 26/min
General: A&O x 3, patient very irritable and physical exam is difficult.
HEENT: PERRLA, EOMI, right TM intact, bulging and red, and no discharge from ears.
Neck: supple, no JVD, no lymphadenopathy, and no thyromegaly.
CVS: S1 and S2, regular, no murmurs, no rubs, and no gallops.
Chest: CTA (B/L), no wheezing, no rales, no ronchi, and no scoliosis.

Abdomen: soft, non-tender, non-distended, +BS, and no organomegaly.
Extremities: no clubbing, no cyanosis, and no edema.
Skin: no rashes

Labs:

CBC, CMP, and lead level.

Results:

Results came back within normal limits

Assessment:

1) Otitis media: Patient has a fever. Infant is very irritable and is constantly playing around with her right ear. Possible cause could be either viral or bacterial.
2) Otitis externa: this was unlikely because pain was not elicited by pulling the pinna of the ear which is classic for Otitis externa.

Plan:

1) Amoxicillin for 10 days.
2) Children's acetaminophen for fevers.

CASE STUDY 2: Strep throat (aka Streptococcal pharyngitis)

Chief Complaint: Sore throat and cough.

History of Present Illness:

Patient is a 7 year old boy who presents to his pediatrician along with his mother with a complaint of a non productive cough and sore throat for approximately 3 days. Patient admits to pain on swallowing. Patient also complains of a runny nose. Mother also adds that he had one bout of fever of approximately 102.3 F one day ago. He denies any nausea, vomiting, headaches, ear pain, and abdominal pain.

Past Medical History: Asthma
Past Surgical History: None
Family History: Non contributory
Medications: Albuterol inhaler PRN
Allergies: NKDA
Birth History: Normal delivery at 38 weeks, no complications Immunizations are all up to date.

Review of Systems:

General: complaining of cough and sore throat.
Skin: negative
HEENT: cough, sore throat, runny nose, and dysphagia.
Neck: negative
Chest: negative
CVS: negative
Gastrointestinal: negative
Urinary: negative
Musculoskeletal: negative
Neurological: negative

Physical Exam:

Vitals: BP: 100/70, Temp100.1 F, Pulse: 88/min, regular. RR: 20/min
General: AAO☐3, normocephalic, and atraumatic.
HEENT: PERRLA, EOMI, no papilledema, TM intact, MMM, enlarged and erythematous tonsils (B/L), mild erythema of posterior pharynx.
Neck: supple, no JVD, no lymphadenopathy, and no thyromegaly.
CVS: S1/S2, regular, no murmurs, no rubs, and no gallops.
Chest: CTA (B/L), no wheezing, no rales, and no ronchi.
Abdomen: soft, non-tender, non-distended, +BS, and no organomegaly.
Extremities: no clubbing, no cyanosis, and no edema.

CNS: CN I2 – 12 intact.
Skin: no rashes

Labs:

CBC, CMP, and Rapid strep test.

Results:

Rapid strep test was positive.

Assessment:

1) Strep throat: the rapid strep test came back positive.

Plan:

1) Amoxicillin 500mg bid for 10 days
2) Follow up after 10 days for reevaluation

Psychiatry

Psychiatry Case 1: Depression

Chief Complaint: 66 years old male presents with the chief complaint of not sleeping well and low level of energy.

History of Present Illness:

Patient says the symptoms started 6 months ago, following his wife's death by the recurrence of breast carcinoma they both have been fighting for 8 years. The low level of energy is constant throughout the day. He also has decreased ability to concentrate, which is now affecting his activities of daily living such as reading news paper or solving crossword puzzle. Over the past 2 months, the patient also reports decrease in appetite and loss of 10 pounds in weight. He says he can't go to sleep at night and lies awake thinking about his wife and children. When he does goes to sleep he usually wakes up early in the morning and cannot go to sleep afterwards. He also reports he has lost interest in things he used to enjoy once and feels depressed and helpless. He denies any suicidal attempts but does report he has thought about it a couple of times.

Current Symptoms:

Duration: Chronic, **Depression**: Yes for 6 months, **Anxiety**: No **Panic attacks**: No, **Psychosis**: No

Past Psychiatric History:

History of Depression No, **History of Anxiety** No, **History of Mania** No, **Suicide attempts/gestures** yes for 6 months

First Visit to a Psychiatrist: 8 years ago when wife was diagnosed with Breast CA.

Last Visit to a Psychiatrist: 4 years ago when the wife's Breast CA was in remission.

Psychiatric Hospital Admission: No

Past Medical History:

Review of Systems: as stated above.
Allergies: PCN
Medications: Metoprolol, Captopril, Thiazide, Spironolactone, and Nitrates.
Family History: HTN, DM, DL, and MI.

Substance Abuse History:

Smoker: No, Illicit Drugs: No, ETOH Yes x 4 drinks qd, Legal History: unremarkable

Physical Exam: BP 133/78   Temp 99 F   P 84   R 18

General: patient is in no acute distress, looks tired with a flat affect, speaks and moves slowly.
HEENT: PERRLA, moist mucus membranes, and NC/AT.
Lungs: CTA B/L, good effort, no ronchi, and no wheezes.
Cardiac: regular rate and rhythm without murmurs, gallops or rubs.
Abdomen soft, non-tender, non-distend, and no HSM.
Extremities: without clubbing, cyanosis or lower edema or calf tenderness.
Skin warm, dry, no rash, no cellulites, and no jaundice.

Mental Status Examination

Appearance: Well-groomed, neat, casual,
Weight: Average
Mood: Depressed
Affect: Sad, Tearful
Speech: Slow, Non-spontaneous
Thought Process: Normal Associations, Circumstantial, Tangential.
Thought Content: Not psychotic
Intelligence: Above Average
Memory: Intact
Insight: Full
Judgment: Good

Assessment/Diagnosis:

1. Major Depressive Disorder induced by significant life change and the death of the spouse

Plan:

1. Cognitive behavioral therapy with appropriate counseling
2. SSRI's: Sertraline or Paroxetine

Psychiatry Case 2: Schizophrenia

Chief Complaint: 30 years old male comes to the hospital with a suicide attempt.

History of Present Illness:

After stabilizing the patient. The patient tells the doctor that he jumped off the building because "voices in his head" told him to do so. His mother tells that for the past 6 years her son has been acting odd and bizarre way. He used to be a straight "A" student with a 4.0 G.P.A. with a full scholarship to the University but had to leave because of his behavior. Since then he has been hospitalized multiple times because of his odd behavior but never for a suicide attempt. Currently the patient lives with his mother but spends most of his time in his room with the doors locked because he tells that he is talking to the radio. The patient has been off his medications for two weeks.

Current Symptoms:

Duration: Acute onset, **Depression**: No, **Anxiety**: No, **Panic attacks**: No **Psychosis**: Yes for 6 years

Past Psychiatric History:

History of Depression: No, **History of Anxiety**: No, **History of Mania** Yes, **Suicide attempts/gestures** Yes, the most recent one.

First Visit to a Psychiatrist: 6 years ago before leaving college.

Last Visit to a Psychiatrist: 4 months ago

Psychiatric Hospital Admission Yes, 3 months ago for running around the streets and threatening neighbors.

Past Medical History:

Review of Systems: as stated above.
Medications: Haldol, Risperidone, Lithium, clonazepam, and seroquel.
Family History: Schizophrenia maternal uncle, OCD sister.
Social/Developmental/Educational: Finished high school, College drop-out.

Substance Abuse History:
Smoker: No, **Illicit Drugs**: Yes, cocaine in college, **ETOH**: No, **Legal History**: No

Physical Exam: BP 129/74   Temp 99.1 F   P 89   RR 20

General: patient is alert but not oriented to time and place.
HEENT: PERRLA and moist mucus membranes.
Lungs: CTA B/L, good effort, no ronchi, and no wheezes.
Cardiac: regular rate and rhythm without murmurs, gallops or rubs.
Abdomen soft, non-tender, non-distend, and no HSM.
Extremities: without clubbing, cyanosis, bilateral tibial fractures and left hip fracture.
Skin: warm, dry, no rash, no cellulitis, no jaundice, and multiple ecchymosed surfaces throughout body.

Mental Status Examination:

Appearance: Unkempt
Weight: Slim
Posture: Stiff
Mood: Angry, Elated
Affect: Confused
Speech: Non-spontaneous, Loose Associations, Flight of Ideas, Incoherent, Pressured
Thought Process: Loose
Thought Content: Hallucinations- auditory, visual, Delusions- Grandiose, psychotic
Intelligence above average
Memory: Abnormal
Insight: None
Judgment: Impaired, Questionable

Assessment/Diagnosis:

1. Schizophrenia (Disorganized type)
2. Suicide attempt

Plan:

1. Hospitalize the patient and transfer to psych ward.
2. Restart the old medications and reassess whether they need to be changed or dosage increased or decreased.
3. If patient gets agitated give 10 mg haldol injection and restraint the patient for 4 hours.
4. Reassess in 24 hours.

Psychiatry Case 3: Bipolar Disorder

Chief Complaint: 31 years old male is brought to the ER by the police because he was running around the streets naked and screaming "I am the next billionaire."

History of Present Illness:

According to the police the man was running around the streets saying the statement mentioned above. His wife who accompanied the patient tells that he is a respectful stock trader and this is the first time she has seen her husband like this in their 3 years of marriage. She tells that 3 weeks ago her husband started to act strangely. He sold all his stocks and said that I have a new business plan and that he is going to open an antique store because that's where the numbers tell him to invest his money. He also went on a business trip to Indiana and spent 25,000 gambling in the casino. He also seemed irritable and distracted and has been working more than usual. She also tells that he has been barely been sleeping 4 hours daily for 2 weeks.

Current Symptoms:

Duration: Acute, **Depression**: No, **Anxiety**: Yes, **Panic attacks**: No, **Psychosis**: Yes

Past Psychiatric History:

History of Depression, **History of Anxiety**: No, **History of Mania** Yes, **Suicide attempts/gestures**: No

First Visit to a Psychiatrist:

Has been seeing a psychiatrist since age of 15.

Last Visit to a Psychiatrist:

3 weeks ago for depression.

Psychiatric Hospital Admission Yes, last admit age of 15 for Mania.

Past Medical History:

Review of Systems: negative
Medications: Lithium, Valproic acid and Venlafaxine.
Family History: OCD and Trichotillomania in maternal aunt.
Social/Developmental/Educational: stock trader and investment banker.

Substance Abuse History:
Smoker Yes 1 pack x 20 years, **Illicit Drugs** Yes, marijuana occasionally, **ETOH** Yes, 2 drinks after work qpm, **Legal History**: arrested for vandalism at age 19.

Physical Exam: BP 110/69   Temp 98.6 F   P 103   R 19

General: alert and oriented to place only.
HEENT: PERRLA and dry mucus membranes.
Lungs: CTA B/L, no ronchi, no wheezes, and good effort.
Cardiac: regular rate and rhythm without murmurs, gallops or rubs.
Abdomen soft, non-tender, non-distend, and no HSM.
Extremities: without clubbing, cyanosis or lower edema or calf tenderness.
Skin: warm, dry, no rash, no cellulitis, and no jaundice.

Mental Status Examination:

Appearance: Well-groomed, Casual
Weight: Overweight
Posture: Normal
Mood: Euthymic, Elated
Affect: Anxious, Confused
Speech: Loose Associations, Flight of Ideas, Incoherent, Pressured
Thought Process: Loose, Circumstantial
Thought Content: Delusions, Grandiose, psychotic
Intelligence Average
Memory: Intact, Abnormal
Insight: Partial
Judgment: Impaired, Questionable

Assessment/Diagnosis:

1. Bipolar Disorder type I (Manic episode)

Plan:

1. Admit the patient to psych ward.
2. Use benzodiazepine or haldol for agitation or psychotic symptoms.
3. Assess medication dosage and restart medication.
4. Discontinue Venlafaxine and start him on an SSRI.
5. Reassess situation in 24 hours.

Obstetrics & Gynecology

CASE STUDY 1: Preeclampsia

Chief Complaint: Epigastric pain and excess weight gain.

History of Present Illness:

This is a 22 year old Spanish-American female who is nulliparous and at 30 week gestation presents to the hospital with prominent epigastric pain for the last few days and excessive weight gain. Patient states that the pain is non-radiating and 8/10 on the pain scale. She had gained 5 pounds in the last month. Patient also complains of severe frontal headaches which started about 2 days ago non-radiating. Patient complains of swelling up of hands and feet and complains that all of a sudden her hands and feet started to shake early this morning that's when she decided to come to the hospital. She also complained of decreased appetite and some shortness of breath with exertion. She denies any chest pain, palpitations. Patient also denies any contractions or vaginal discharges.

Past Medical History: urinary tract infection 3 months ago for which she was treated.
Past Surgical History: none
Family History: non contributory
Medications: prenatal vitamins
Allergies: NKDA
Social History: no smoking, no alcohol use, no drug abuse, patient is married, and lives with her husband.
OB/GYN: Nulligravida and periods were regular before pregnancy.

Review of systems:

General: excessive weight gain for the past month and decreased appetite.
CVS: negative
Chest: shortness of breath with exertion.
Gastrointestinal: epigastric pain past 2 days non-radiating.
Urinary: decreased amount of urination.
Musculoskeletal: shaking of arms and legs.
Extremities: swelling of legs and hands.
Neurological: frontal headaches
Psychiatric: negative

Physical Exam:

Vitals: BP 182/112, HR 95, Temp 99.0 F, RR 22 O2 Sat 95% on RA.
General: alert and oriented to person, place, and time.
HEENT: atraumatic, pupils equal round and reactive to light and accommodation, extra ocular muscles intact, throat moist mucosa, swelling of the eyes, and pallor conjunctivae.
Neck: supple, no thyromegaly, and no masses.

Lungs: bilateral rales at the bases, no wheezes, and no crackles.
CVS: S1 and S2 audible, grade 2/6 systolic apical murmur, and sinus tachycardia.
Abdomen: soft, + tenderness with palpation in the epigastric region, mild distension + bowel sounds, + tenderness and palpable liver 3 fingers below the right costal margin.
Extremities: 2+ edema of the legs and hands and no rashes observed.
Neuro: CN 2-12 intact and no gross neurological deficits.

Labs:

1. CBC, CMP, U/A, Urine culture, 24 hour urine collection for proteins.
2. ULTRASOUND of the abdomen

Results: labs came back normal at pregnancy

1. Urine showed 5 grams of protein over a period of 24 hours which suggested the diagnosis of preeclampsia.
2. Ultrasound did not show any abnormality.

Assessment:

1) Severe preeclampsia: Patient's urine showed 5 grams of proteinuria over the period of 24 hours. Patient's clinical signs and symptoms also correlated with the diagnosis.
2) Eclampsia: was a possibility but there was no evidence of seizures.
3) Pregnancy induced hypertension: was excluded because patient will not have proteinuria with this.
4) HELLP syndrome: Patient did not have this syndrome because the liver enzymes and platelets were not elevated significantly.

Plan:

1) Patient was admitted.
2) Patient was monitored for the next 24 hours.
3) IV magnesium sulfate was given to prevent convulsions.
4) Fetal stress testing was done to see if there was fetal jeopardy.

Abbreviations:

1. A&O – alert and oriented
2. AST - aspartate aminotransferase
3. ALT – alinine aminotransferase
4. ALK Phos – alkaline phosphatase
5. ABG – arterial blood gas

6. BP – blood pressure
7. B/L – bilateral
8. BS – bowel sounds
9. BNP – beta - natriuretic peptide
10. BGS – blood glucose

11. CXR – chest X-ray
12. CN – cranial nerves
13. CTA – clear to auscultation
14. CBC – complete blood count
15. CMP – comprehensive metabolic profile
16. CVS – cardiovascular system
17. CC – chief complaint
18. CT – computed tomography

19. DM – diabetes mellitus
20. DL – Dyslipidemia

21. ETOH – ethanol, alcohol
22. ER – emergency room
23. EKG – electrocardiogram
24. ECHO – echocardiography
25. EOMI – extra ocular muscles intact

26. FSH – follicle stimulating hormone
27. FOBT – fecal occult blood test

28. GERD – gastro esophageal reflux disease

29. HEENT – head, eyes, ears, nose, and throat
30. HR – heart rate
31. HCTZ – hydrochlorothiazide
32. HELLP syndrome – hemolysis, elevated liver enzymes, and low platelets
33. HPI – history of present illness
34. HTN – hypertension
35. HSM – hepatosplenomegaly

36. IV – intravenous

37. JVD – jugular venous distention

38. LH – luteinizing hormone
39. LMP = last menstrual period

40. MMM – moist mucous membranes
41. MI – myocardial infarction
42. MVA – motor vehicle accident
43. MVT – multivitamin

44. N/V/D/C – nausea, vomiting, diarrhea, constipation
45. NPO – nil per os
46. NC/AT – normocephalic, Atraumatic
47. NKDA – no known diagnosed allergies
48. NKA – no known allergies

49. O2 Sat – oxygen saturation
50. OCD – obsessive compulsive disorder
51. OTC – over the counter

52. PERRLA – pupils equal round and reactive to light and accommodation
53. PRN – as needed
54. PCN – penicillin
55. PMH – past medical history
56. PSH – past surgical history
57. PSA – prostate-specific antigen
58. PE – pulmonary embolus
59. PPD – pack per day
60. Pt – patient
61. PPI – proton pump inhibitors

62. Qd – once daily

63. RR – respiratory rate
64. RA – room air
65. RRR - regular rate and rhythm

66. SSRI – selective serotonin reuptake inhibitors
67. SOB – shortness of breath
68. STD – sexually transmitted disease

69. Temp – temperature
70. TM – tympanic membrane
71. TSH – thyroid stimulating hormone

72. UTI – urinary tract infection
73. URI – upper respiratory tract infection
74. U/S – ultrasound
75. UA – urinalysis

76. V/Q scan – ventilation perfusion scan

77. WBC – white blood count
78. WNL – within normal limits

79. + / - - Positive or Negative

www.ingramcontent.com/pod-product-compliance
Lightning Source LLC
Chambersburg PA
CBHW081850170526
45167CB00007B/2954